Preface

Nuclear energy has been a very grey area in many people's lives. This is despite the fact that nuclear energy is continually playing a significant role in the world and in the coming future. In light of this situation, it is important to ease the young generation into acquiring some basic knowledge in this field. This is a fun and interactive way to learn what at times is considered to be a distant and complex field. The book uses common words to replace complicated vocabulary. There are also countless easy-to-relate examples that the reader can benefit from.

Contents

Class 1: Introduction to nuclear science ... 3
Class 2. What is Nuclear Fission? ... 6
Class 3: Nuclear Fusion ... 8
Class 4: How do we use nuclear energy ... 10
Class 5: Other Applications of Nuclear Energy ... 12
Class 6: Stakeholders involved in the nuclear industry ... 14
Class 7: Nuclear safety measures I ... 17
Class 8: Nuclear safety measures II ... 19
Class 9. How does nuclear energy compare against other energy sources? ... 21
Class 10: The future of nuclear energy in the world ... 24

Nuclear Energy Science
Class 1: Introduction to nuclear science

What is nuclear energy? Well, I don't think that is where we should begin. The story starts from the small particles that make up everything in this universe. These are called atoms. It is from the atoms, that we get the smaller particles called protons, neutrons and electrons.

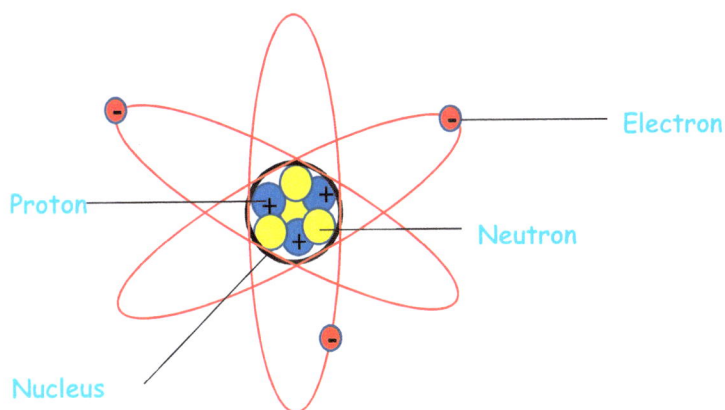

Now as we can see from the diagram, the protons are positive in charge, and electrons are negative in charge. Just like a battery has a positive side, and a negative side. However, the neutrons are neutral, meaning they do not carry any charge. In general, the nucleus is positive in charge, as the positive charge of the protons are unopposed by the neutrons. In nuclear science we will focus more on the neutron for reasons that I will explain later. Another interesting fact, is that the neutron is slightly heavier than the other two type of particles.

Why Is the Neutron Important? The electrons revolve around the atom's nucleus, which is made of neutrons and protons. There is a force that holds the electrons in place within the atom. This is called the electrostatic force of attraction. Now this is not so difficult to understand, it basically means that particles of the opposite charge attract each other. Just like a magnet's south pole is attracted to another magnet's north pole. The opposite of this force, is the electrostatic force of repulsion, that makes the like charges repel. Like a north pole repelling another pole.

The theory in nuclear science, is to use the neutron to create a series of reactions within special materials in order to produce heat. I know what you are thinking now "wow, that sentence sounds quite difficult to understand". But let us dissect it bit by bit. Remember we said that the neutron is neutral and also has a higher mass than the other particles. Hence, it is used to strike an atom to create smaller atoms, extra neutrons and heat energy. The fact that the neutron is neutral, it is able to hit the atoms without being repelled by either the protons or neutrons. This brings us to a very important subtopic

Introduction Practice Questions section to help you revise class 1.

1) Match each of the particles with their respective charges

Protons	Neutral
Neutrons	Positive
Electrons	Positive
Nucleus	Negative

2) Electrical charges of the same kind attract each other. Is this statement true or false?
    A. True   B. False

3) Which two particles make up the nucleus in an atom?
    A. Proton and electron
    B. Electron and neutron
    C. Proton and neutron
    D. None of the above

4) Which of the following is not produced when an atom is split?
    A. Smaller atoms
    B. Other neutrons
    C. Sound
    D. Heat

5) Why is the neutron much more important in nuclear science?
    A. It is heavy and negative
    B. It is heavy and neutral
    C. It is heavy and positive
    D. None of the above

Class 2. What is Nuclear Fission?

So far, we have learnt quite a lot, and now you can tell all your buddies how much you know about nuclear science. But before you do that, we have a little more to cover. From, the previous class we learned about the structure of the atom and why the nuclear is important to nuclear science. We touched on the splitting of atoms in order to produce, more smaller atoms, more neutrons and heat energy. This is commonly known as nuclear fission

The image below offers a better understanding of the process.

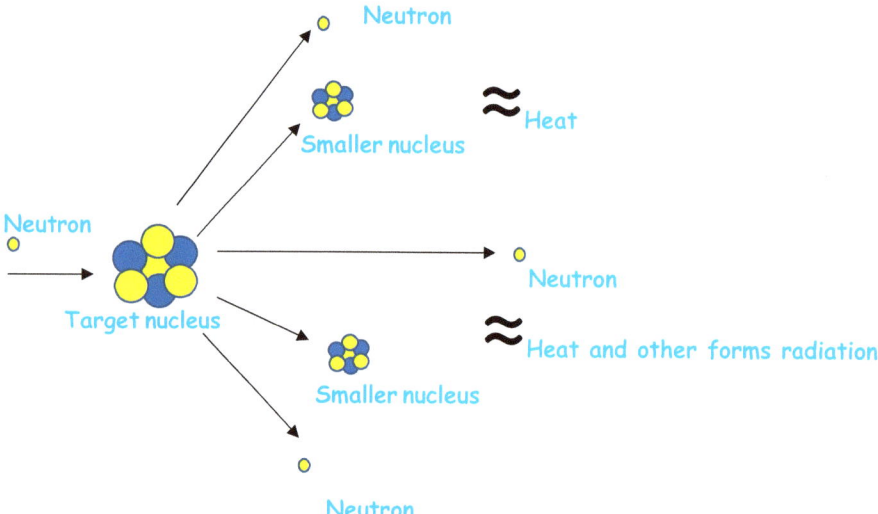

As we look at this image, we should remember an example of the small snowball rolling downhill. The snowball starts from the top while being very small. However, as it continues to roll downhill, it gathers more snow along the way and hence becomes bigger and bigger. The nuclear fission process is very similar to the snowball effect.

This is because, as the neutron hits the atom's nucleus, two or three extra neutrons are created. These new neutrons in turn strike other atoms producing more neutrons, atoms and more heat. Not all materials are able to undergo fission, only special materials such as uranium metal undergo this process. These are called fissionable materials. The ability to release the nuclear energy continually is called radioactivity.

As we can see in this trend, the heat will continue increasing, and that is where nuclear science starts to become useful to us human beings. A little later, we will discuss how we control and use this heat and even convert it to other forms of energy.

Now in nuclear science, we can use another method to get this heat. Instead of hitting the nucleus with the neutron, we now combine two nuclei to form a bigger nucleus, neutrons and produce heat. This is called nuclear fusion. However, we will reserve this new concept for Class 3. It is time to do some exercise.

Nuclear Fission Practice Questions to help you revise class 2.

1) Which of the following particles is used to start a nuclear fission process?

   A. Nucleus
   B. Neutron
   C. Electron
   D. Proton

2) What is the term used to refer to the splitting of an atom to produce smaller atoms and neutrons?

   A. Nuclear Fission   B. Nuclear Fusion

3) There is more than one way of creating nuclear energy. Is this true or false

   A. Nuclear Fission   B. Nuclear Fusion

4) Which of the following is a fissionable material?

   A. Uranium
   B. Oxygen
   C. Iron
   D. Copper

5) What form of energy is mainly used from the fission process?

   A. Sound
   B. Light
   C. Heat
   D. Vibration

Class 3: Nuclear Fusion

From the last class, we learned how certain special metals such as uranium are used in nuclear fission to produce heat energy. We discussed the nuclear fission process and the introduced the term fissionable material. In this class we will look at another method that is used in the nuclear science field.

Nuclear fusion is almost the opposite of nuclear fission. Instead of splitting the nucleus, now we combine two nuclei to form one bigger nucleus and release energy at the same time. Again, just like in nuclear fission, not all materials or elements are suitable to use in nuclear fusion.

One common element used in nuclear fusion is hydrogen. The hydrogen nuclei are combined to form a helium nucleus and energy. This may sound so complicated and new, but surprisingly, nuclear fusion happens every day in our life. How does it happen you ask? Well the sun you see on a daily basis is an example of nuclear fusion happening. Incredible isn't it? For the sun to give us heat, it converts the hydrogen to helium as explained above.

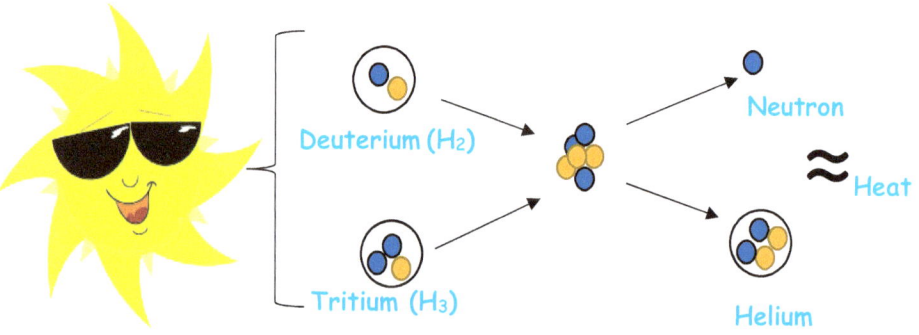

The image above describes the nuclear fusion process, just as it happens in the sun and in research facilities. Deuterium is a hydrogen atom that has one neutron and one proton but tritium is a hydrogen atom that has 2 neutrons and one proton.

Nuclear fusion unlike nuclear fission, is still under research in order to make it more useful to us human beings. This is because the energy and temperature reached in a nuclear fusion process are very high. More research is needed in order to develop strong materials to handle these conditions.

Nuclear Fusion Practice Questions to help you revise class 3.

1) Nuclear fusion is the joining of two nuclei to form a big nucleus and release energy.
   A. True   B. False
2) Which element is commonly used for nuclear fusion?
   A. Helium
   B. Uranium
   C. Hydrogen
   D. Oxygen
3) Which of the following body is powered by nuclear fusion?
   A. Planets
   B. The sun
   C. The moon
   D. None of the above
4) Which element is produced when two hydrogen nuclei combine?
   A. Helium
   B. Uranium
   C. Hydrogen
   D. Oxygen
5) Which nuclear process produces more heat energy
   A. Nuclear Fission   B. Nuclear Fusion

Class 4: How do we use nuclear energy

From the previous class, we learned about nuclear fusion, we learned how different it was from nuclear fission. We saw how the sun uses this process to give heat, and what elements are involved. Now, that you are well versed with nuclear energy production methods, let us look at the applications of nuclear energy today and in the near future.

Some of the common uses of nuclear energy are the generation of electricity, district heating, water purification, medical applications, security scanning equipment. Let's look at the major one:

Production of electricity:

The heat developed in nuclear a fission reaction, is used to heat water. The water is converted into steam. The steam is then used to turn a turbine, which then drives a generator. The generator converts the motion to electricity. The whole process takes place inside a nuclear power plant.

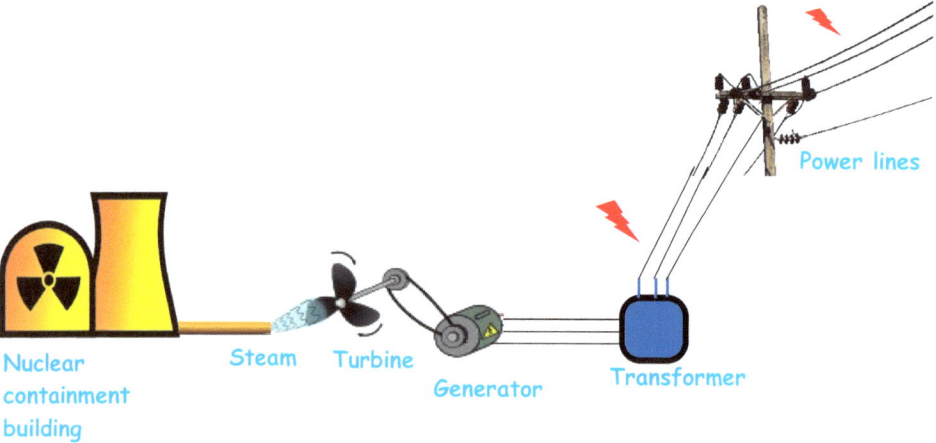

The image above describes the process in which nuclear heat is converted to electricity. In currently there is a lot of research to improve the method and make use of other fluids apart from steam. However, the rest of the energy conversion cycle is the same. The turbine converts the steam to mechanical rotation, the generator picks up this rotation and produces electricity. The transformer, increases the amount of electricity produced and the electricity is carried by the power lines.

Practice questions on how to use nuclear energy

1) What is heated to produce steam in the nuclear building?
   A. Water
   B. Uranium
   C. Oil
   D. None of the above
2) Which of the following equipment is used to convert steam to mechanical rotation?
   A. Generator
   B. Turbine
   C. Transformer
   D. Power line
3) Which of the following equipment is used to convert mechanical rotation to electricity?
   A. Generator
   B. Turbine
   C. Transformer
   D. Power line
4) Which of the following equipment is used increase the amount of electricity so that it is transmitted?
   A. Generator
   B. Turbine
   C. Transformer
   D. Power line
5) Steam production is the nuclear plant is the only technology that can be used to produce electricity. Is this statement true or false?
   A. True   B. False

Class 5: Other Applications of Nuclear Energy

From the previous class, we saw how nuclear energy is converted into electricity. We saw the importance of certain equipment such as generator, turbine, and transformer. In this class we will look into other applications that make use of nuclear power:

Application	Description
Public area heating	The hot water produced by the nuclear power plant is put into steam pipes. The heated water is distributed to homes and companies during winter.
Purification of water	The heat energy and electrical energy from a nuclear power plant are both used to purify water. This is mainly for seawater whose impurities and salts are removed in the process. The water is then used for public consumption.
Security scanners	The radiation from some the radioactive materials such as Cobalt is used. The radiation is able to penetrate materials and create an image displaying the contents inside containers or bags. This is common in security checkup areas such as in airports
Medical applications	Radiation is used in scanning equipment such as MRI and CT scanners in hospitals. X-ray machines are also used to create internal body images.
Military applications	The nuclear energy is used to power military equipment such as air-craft carriers, submarines and some space shuttles.

As we can see the nuclear energy is very important in many areas in our lives. The uses of nuclear science are continually increasing as more research is done in the field. It is a very promising field today and in the future.

Practice questions on Other Applications of Nuclear Energy

1) What is used in public area heating can be done by use of nuclear power plants? Is this statement true or false?
   A. True
   B. False
2) Which of the following applications are considered as medical applications for nuclear energy?
   A. Powering a submarine
   B. MRI scanner
   C. Customs security check
   D. Public area heating
3) In which of the following vessels would you find a small nuclear plant?
   A. Public bus
   B. Military ship
   C. Train
   D. Airplane
4) What form of energy is used in X ray scanners?
   A. Nuclear radiation
   B. Sound
   C. Heat
   D. Light
5) Give an example of a radioactive element that can be used in the security scanners.
   A. Potassium
   B. Cobalt
   C. Carbon
   D. Copper

Class 6: Stakeholders involved in the nuclear industry

From the last class we dived deeper into more applications of nuclear energy. We saw how the nuclear energy is utilized in medical fields, water purification, public area heating, scanning equipment and military vessels. In this section we shall look at the bodies that are involved in the nuclear industry.

Unlike many other industries, the nuclear field, has many interconnected bodies. It is not a stand-alone field. There are many companies and organizations that work together in order to make the nuclear industry tick. The image below shows the integration of the whole nuclear industry with different stakeholders or interested parties.

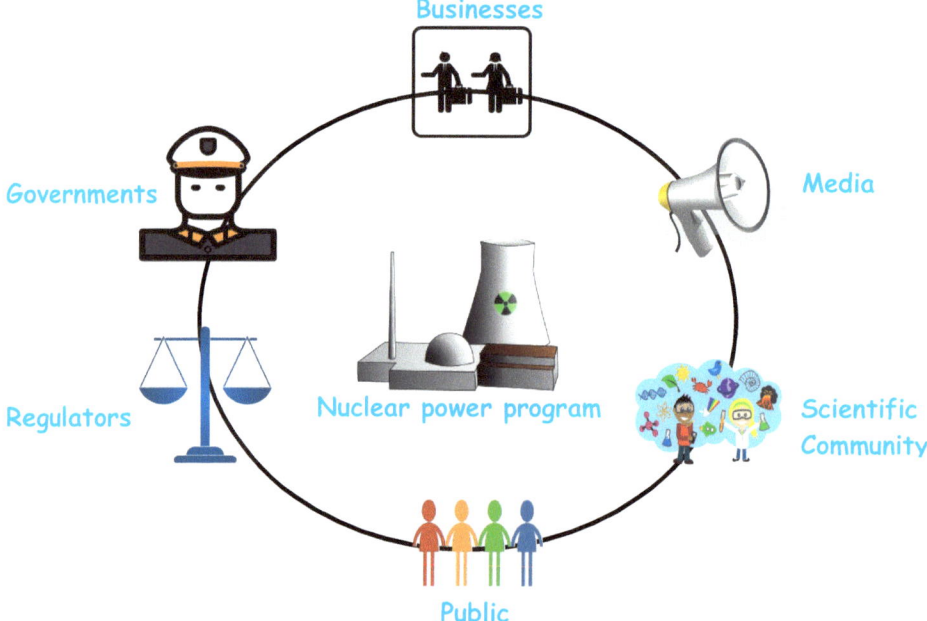

The businesses refer to corporation and organizations such as uranium mining companies, manufacturers of nuclear equipment, designers of power plants, operators of power plants and nuclear waste handling facilities. These organizations interact amongst themselves and with other players in the industry.

The governments are involved in setting up local policies for operations of the nuclear power plants in their own nations. They are also in charge at times in funding the nuclear programs and hence they are an important stakeholder. The regulators on the other hand create the international laws which govern the use of nuclear energy in the world. An example of the regulators is the International Atomic Energy Agency (IAEA). The countries wishing to adopt nuclear energy global integration must be signatories to this agency.

The public are inclusive of ordinary people either living near the power plants, or being served by the products of the nuclear plant. Their main concern is safety of nuclear energy and hence their concerns must be addressed. The other concern they may have is the cost of implementing the plant within the country or region.

The scientific community is very important too. This is inclusive of universities and nuclear research facilities. It is from these organizations that we get inventions and innovations on nuclear energy fields. Finally, the media plays a fundamental role in informing the public masses on issues that pertain the nuclear industry. Without an effective media, the public would be blind to what the other industrial players are up to in terms of nuclear development.

Practice questions on Stakeholders involved in the nuclear industry

1) Which of the following organizations is responsible for funding nuclear power programs in countries?
   A. Media
   B. Government
   C. Regulators
   D. Scientific community
2) Which of the following organizations is responsible for the setting up international laws of governing nuclear programs
   A. Businesses
   B. Government
   C. Regulators
   D. Scientific community
3) Which organization is an example of nuclear power regulators?

A. World Health Organization
   B. Food Agricultural Organization
   C. United Nations
   D. International Atomic Energy Agency
4) What is the role of the media?
   A. Funding nuclear programs
   B. Informing public on the progress of nuclear programs
   C. Form laws to govern nuclear programs
   D. Invent new nuclear technologies
5) Which organization is responsible to create innovations and inventions on nuclear power plants technologies?
   A. Media
   B. Government
   C. Regulators
   D. Scientific community

Class 7: Nuclear safety measures I

Well done for having come this far, you have learned so much within the nuclear field, and guess what? There is more cool stuff coming your way in the coming topics. In the last class we covered the stakeholders who are involved in the nuclear industry. In this class we will cover safety issues with regards to nuclear energy safety. Due to the huge potential of nuclear energy and the radiation released we need to take care when handling nuclear related equipment.

We will look into the measures that are inherently built on the reactor power plant. This is in order to safeguard it against emission of radiation and control heat production in the plant

Radiation containment. The nuclear power plant is designed to ensure, that the radioactivity released during operation and even during accidents is contained. Radiation is harmful and hence nuclear operators have containment measures in place. Some of them are, nuclear fuel cladding ("coating"), reactor vessel protection barrier and containment building barriers. The containment barrier also protects the plant from external impact such as airplane strikes.

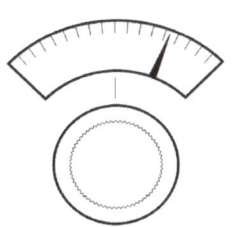

Control of radioactivity. A nuclear power plant's radioactivity must be controlled. This prevents rampant increase in radioactivity. This is why the nuclear power plant does not explode like a nuclear bomb does. The control measures are also used in starting, speeding up, slowing down and shutting down a nuclear power plant. Just like the gas pedal and brake pedal in a vehicle. Common control measures in a nuclear power plant are: control rods and use of boron chemical.

Heat removal. As we saw in our previous topics, heat is the primary product of the nuclear plant. Hence this heat must be "removed" from the reactor and passed to the steam

production process. If heat is allowed to build up without removal, this would damage the reactor internal components such as the nuclear uranium fuel.

Practice questions on nuclear safety measures II

1) Which of the following is not used as radiation shield in a nuclear power plant?
   A. Cladding coating of the fuel
   B. Reactor vessel barrier
   C. Boron chemical
   D. Containment building
2) Which of the following is used in controlling the reactivity rate of a nuclear reactor?
   A. Control rod
   B. Dosemeter
   C. Pump
   D. Containment building
3) Which of the following structure protects the nuclear plant from external impacts
   A. Cladding coating of the fuel
   B. Reactor vessel barrier
   C. Containment building
   D. None of the above
4) What is the most severe impact of overheating of a nuclear plant?
   A. Production of steam
   B. Melting of reactor fuel
   C. Slowing down of the reactor
   D. Shutting down of the reactor
5) What is the immediate primary product of a nuclear power plant?
   A. Purified water
   B. Electricity
   C. Steam
   D. Heat

Class 8: Nuclear safety measures II

The last class involved the studying of nuclear safety measures used in the nuclear plant to protect against radiation emission, overheating and external impacts.

In this class we will look into the secondary measures of safety that extend from the nuclear reactors.

Worker's safety Nuclear facility workers need to be protected from the harmful radioactive materials inside a power plant. Operators ensure that workers wear personal protection equipment and use special tools to prevent them from being exposed to radiation.

They wear nuclear dosage meters to detect the amount of radiation they are exposed to within a particular period.

Spent nuclear fuel disposal Nuclear waste is mainly the spent nuclear fuel that is recovered from a nuclear plant. The spent fuel is still radioactive and hence it should be handled carefully. It is mainly stored in deep underground areas where there is no access to the public. At times the spent fuel can be re-processed to form new fuel.

Nuclear safeguard policies. Nuclear regulators are in charge of developing safeguard policies that regulate the handling of nuclear material. They establish measures to ensure tracking of nuclear and radioactive materials for all the nuclear facilities. This is to prevent the conversion of civilian nuclear material to military grade nuclear material.

Practice questions on nuclear safety measures II

1) What device measures the radioactivity that the nuclear facility workers are exposed to?
   A. Control rod
   B. Dosemeter
   C. Pump
   D. Containment building
2) What is the main way of disposing spent fuel from nuclear reactors?
   A. Stored in underground facilities
   B. Burning it
   C. Reprocessing it
   D. None of the above
3) Nuclear facilities workers need to wear personal protection equipment when handling nuclear material.
   A. True
   B. False
4) Why is it important to monitor the movement of spent nuclear fuel?
   A. Prevent conversion to military-grade material
   B. To prevent theft
   C. To control the global prices
   D. None of the above
5) Burying spent fuel in underground facilities is the only way of dealing with spent fuel. Is this statement true or false?
   A. True
   B. False

Class 9. How does nuclear energy compare against other energy sources?

In the previous topic we looked into the nuclear safety features such as handling of spent fuel, workers' protection material and regulatory laws. In this class we will evaluate how nuclear power compares to other common conventional energy methods.

Coal power -the coal power plant is the main competitor to nuclear especially in generation of huge amounts of electrical power. A coal power station uses coal to heat water to produce steam which is used as the steam in a nuclear plant.

The benefit it has over nuclear power is that it has a lower cost of capital when starting it. However nuclear power is cleaner as it does not have polluting gases like the ones present in coal power plants. More to that, the coal mines create a lot of destruction to the environment during the extraction process.

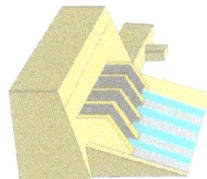

Hydropower- hydropower is the use of water to turn a turbine instead of steam. To do this, a river is trapped from flowing using a huge wall and the water is then channeled at high pressure to the turbine.

The advantage of hydropower over nuclear power is that, hydropower does not need "refueling" like a nuclear power. The advantage of nuclear power is that it does not rely on weather patterns like hydro energy. Plus, hydro energy involves disruption of larger area of land and resettlement of people.

Oil and gas- this is commonly known as fossil fuels. They can be used directly in combustion engines to create electricity without going through the steam cycle.

Initial costs of these plants are lower than nuclear power plants however, the costs of fuel and running costs are higher in fossil fuel plants. More to that, these plants, just like coal plants emit some polluting gases to the environment.

Solar power – Solar energy is the use of the sun's radiation on special panels to create electricity. The advantage of this form of energy is that it does not require refueling like nuclear energy.

The disadvantage of this form of energy is that it is not reliable, it only works when the sun's light is present. Other than that, the amount of land needed to set the panels is very large for the same amount of energy that can be obtained from a nuclear plant.

Wind power - The wind power, uses wind to turn a turbine instead of steam. The advantage of this form of energy is that, its cheap to maintain and does not need refuelling. However, just like solar energy, this form of energy is not reliable, the wind blows only when it is present.

More to that, the amount of land needed to set the panels is very large for the same amount of energy that can be obtained from a nuclear plant.

Practice questions for how does nuclear energy compare against other energy sources

1) Why is nuclear power considered cleaner than coal power?
   A. It uses steam to turn the turbines
   B. It uses uranium.
   C. It does not have pollutant gases
   D. None of the above
2) What is used to drive the turbine in hydropower plants?
   A. Wind
   B. Steam
   C. Engine
   D. Water
3) Oil and gas power plants are more expensive to run than nuclear power plants.
   A. false
   B. True
4) What is the disadvantage of solar power as compared to nuclear power?
   A. It is unreliable

B. It is expensive
C. It has pollutant emissions
D. None of the above

5) What is the disadvantage of wind power as compared to nuclear power?
A. It emits harmful polluting gases.
B. It is expensive more than a nuclear power plant
C. Land required for a wind farm is very large for the same power as a nuclear power plant
D. None of the above

Class 10: The future of nuclear energy in the world

In the last class we studied the other forms energy and compared them to nuclear energy. In this class we will consider what the future holds for nuclear energy. There is a lot of research underway in order to find new ways to use nuclear energy. As we have seen, nuclear energy is safe, clean and relatively more reliable.

Improvement in safety features- as the nuclear communities continue learning on new technologies, safety features of the plants' increase. It means that nuclear plants in the near future can easily be integrated in residential areas. Small modular reactors are an example of reactors, that can be placed close to human settlements

New reactor types-the need to save on uranium deposits, and the need to have more efficient plants has led to research to develop new type of reactors. These use different technologies such as helium gas, molten salt or molten lead to cool the core. These are able to make use of spent fuel from conventional nuclear reactors and also have better efficiency.

Regulations-due to nuclear incidents, the laws involving handling of nuclear material continue to improve. Stringent measures are put in place to ensure that the nuclear industry operates without hiccups.

Advanced applications- the market for applying nuclear energy continue growing tremendously. The use of nuclear energy in space shuttle programs, military vessels, and in medical research promises to have a bright future for the nuclear industry.

Practice questions on the future of nuclear energy in the world

1) What kind of plant can be placed near residential areas?

    A. Coal plant
    B. Hydro plant
    C. Small modular reactor plant
    D. None of the above

2) Which of the following elements cannot replace water in the cooling of a nuclear reactor

    A. Molten lead
    B. Helium
    C. Oxygen
    D. Molten salt

3) Nuclear regulations are constant and never change with time

    A. False
    B. True

4) There is a lot of research underway in order to find new ways to use nuclear energy?

    A. False
    B. True

5) Which of the following is not a currently researched areas for application of nuclear power?

    A. Space shuttle program
    B. Medical applications
    C. Military vessels
    D. Airplanes

www.ingramcontent.com/pod-product-compliance
Lightning Source LLC
Chambersburg PA
CBHW040350220526
45473CB00009B/2834